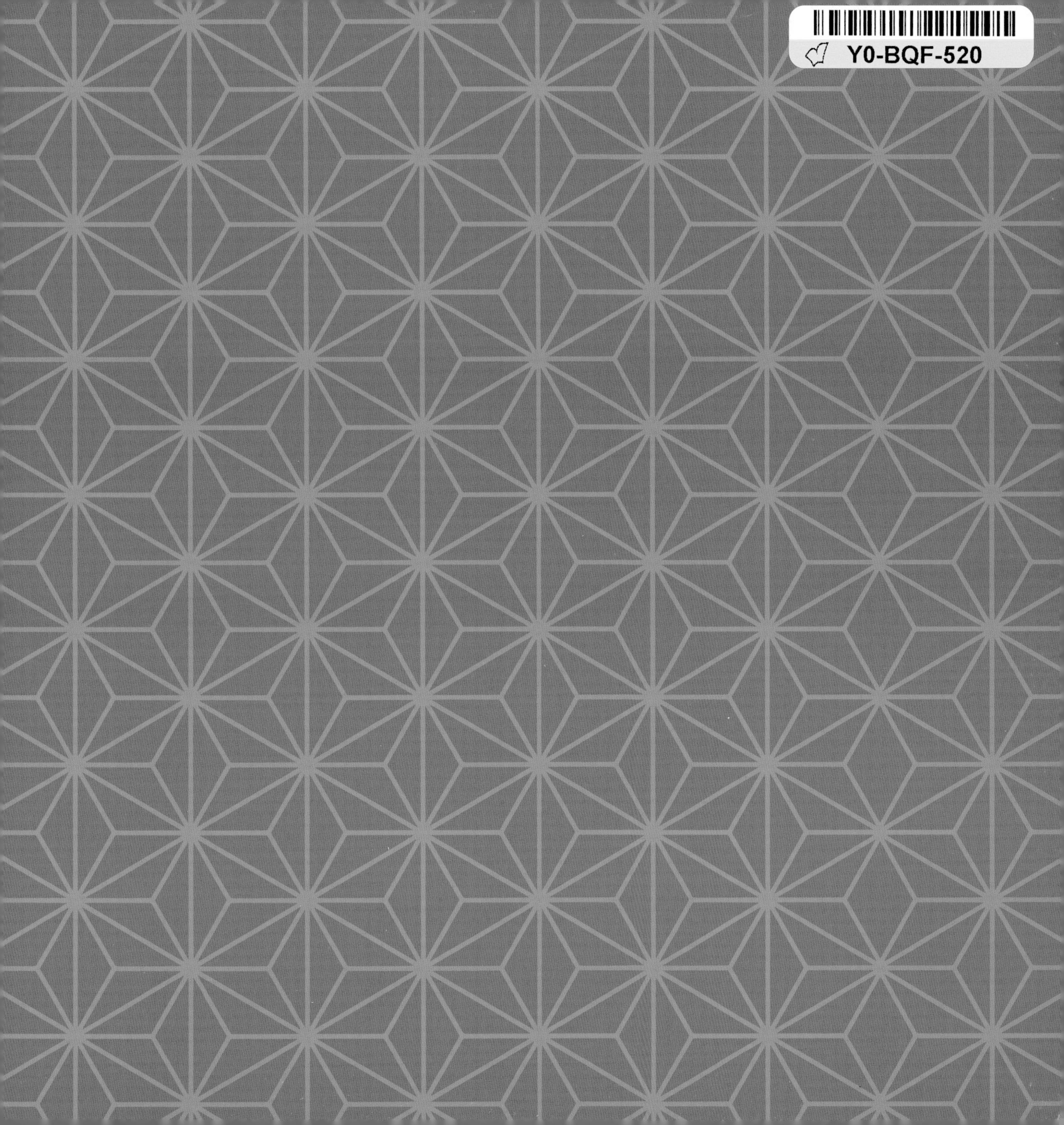
Y0-BQF-520

Text and illustration copyright © 2022 by KiwiCo, Inc.

All rights reserved. No part of this book may be reproduced in any form without written permission from the publisher.

ISBN: 978-1-956599-00-8

Library of Congress Control Number: 2021949978

Manufactured in China.

10 9 8 7 6 5 4 3 2 1

KiwiCo, Inc.
140 East Dana Street
Mountain View, CA 94041

KiwiCo.com

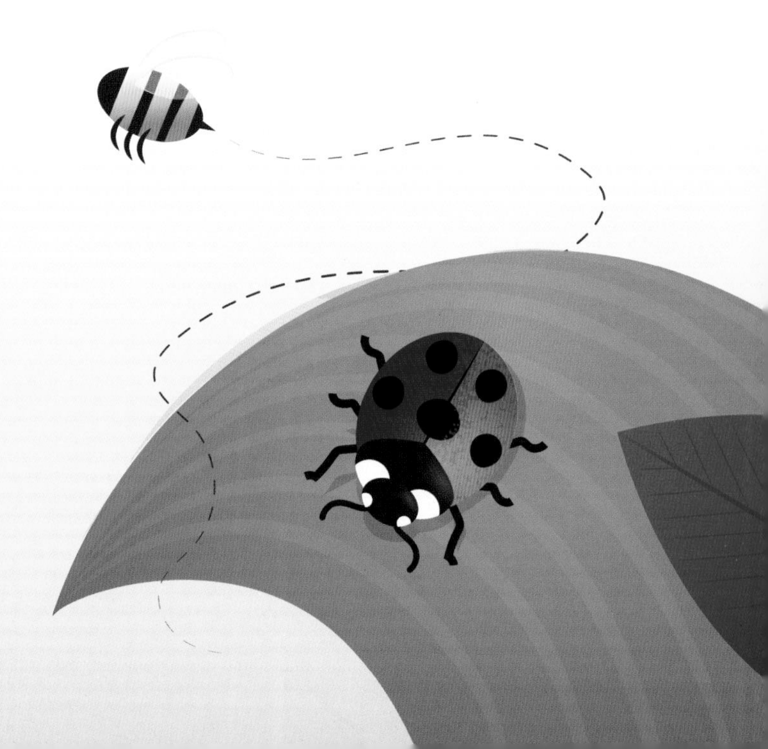

OTTO's GRAND SYMMETRY TOUR

JILLIAN SCHRAGER

KiwiCo Press

Hi there, fellow adventurer!

I'm OTTO, your host on the Grand Symmetry Tour: a journey in search of the most amazing, fascinating, awe-inspiring symmetrical stuff!

At every stop, I'll shrink us down to the best size to see all the symmetrical details. Then I'll need *your* help to get us to the next symmetrical wonder by completing a fun challenge.

Let's hit the road!

Have you ever noticed that your face is the same on both sides? You have two eyes, two ears, and two nostrils.

Drawing a line down the center of your head divides your face into two perfect halves.

Things that are the same on both sides have **bilateral symmetry**.

If you look at a flower, you'll find another type of symmetry. See how the petals come out of the center in the same way all around? This is called **rotational symmetry.**

Got it? Great! You're almost ready to start the tour . . .

But first, let's talk about things that *aren't* symmetrical!

Say hello to these two shrimps.

One has bilateral symmetry, and the other, a pistol shrimp, has a very big claw on only one side. This is called **asymmetry.**

Things with symmetry are called **symmetrical.** Things with asymmetry are called **asymmetrical.**

These cookies are circles, so you might think they're symmetrical. But take a closer look at the chocolate chunks.

Even if we break the cookie into perfect pieces, the chocolate chips won't be in exactly the same places. So the chocolate makes the cookies asymmetrical. (And yummy.)

Now you know everything you need to know for the Grand Symmetry Tour!

Our first stop on the Grand Symmetry Tour is a place you're probably very familiar with . . .

The kitchen is full of tasty, symmetrical treats. Nature makes symmetry in fruits and vegetables, and we sometimes make symmetry in the things we cook and bake.

On this page, find six types of food with bilateral symmetry.

Hint: a few are trying to fool you!

Answer: lemon, onion, tomato, avocado, pepper, cupcake

Let's take a look at some of my favorite round foods!

Citrus fruits, such as oranges and grapefruits, grow outward from a single bud. Because they grow at the same rate on all sides, they turn into symmetrical spheres.

The inside of an orange looks like a fruity, juicy bike wheel. And if you spin an orange like a wheel, you'll see that it looks the same as you turn it. (Spin the book to give it a try!) That tells you oranges have rotational symmetry.

What if an orange were all
peel on one side and all
wedges on the other?
How would you eat it?
Could you carry it as easily?

Onions have rotational symmetry too, but they don't have sections. They grow outward from an onion stem in circular rings that get bigger and bigger.

Math always makes me hungry, so let's look at these pies.

Each pie is symmetrical in a different way. Look at the holes in the crust. The green pie has five holes, so you can cut it into five equal slices. The red one has eight holes, so you can cut it into four or eight slices.

What about the purple pie?

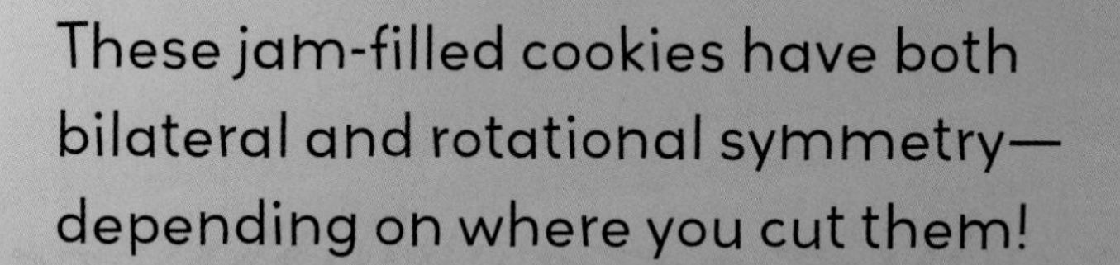

These jam-filled cookies have both bilateral and rotational symmetry—depending on where you cut them!

What if pans were lumpy and bumpy instead of round or rectangular?

How would you decorate these cupcakes? How would you make them symmetrical or asymmetrical?

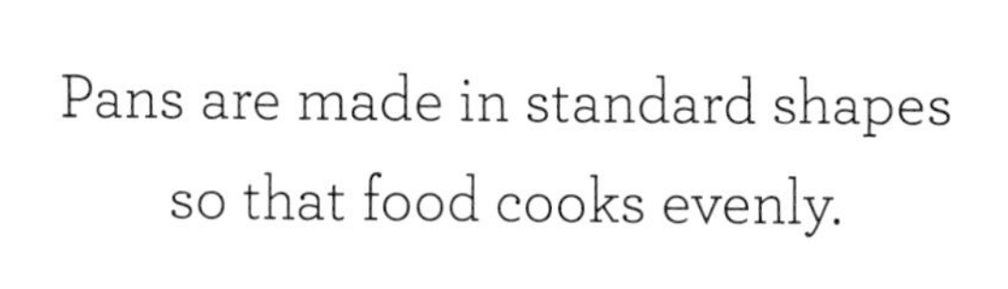

Pans are made in standard shapes so that food cooks evenly.

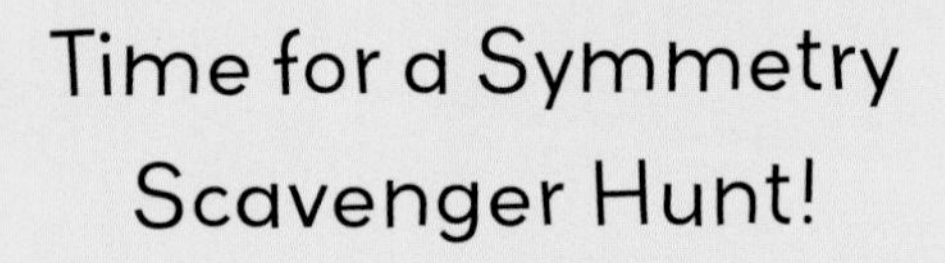

Time for a Symmetry Scavenger Hunt!

To move on to the next stop, look around your room for . . .

- One object with bilateral symmetry
- One object with rotational symmetry
- One object with asymmetry

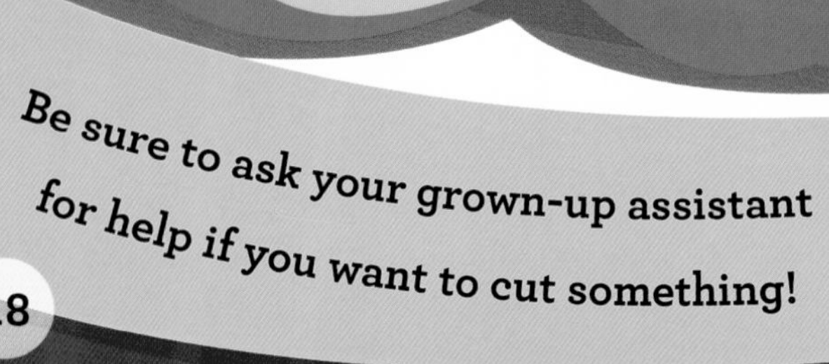

Be sure to ask your grown-up assistant for help if you want to cut something!

Ready, set, hunt!

On to our next stop . . .
. . . the garden!
Gardens are great places to look for symmetry. I love the symmetry of flowers, and so do birds and bees!

Symmetry isn't just for pretty flowers. Check out this orb spider's web—but don't get caught! Doesn't it look like a kaleidoscope?

Not all spiders weave symmetrical webs. This cob spider's tangle web has asymmetry instead.

If you were a spider, what type of web would you make? Try drawing your own!

One day, I almost fell asleep in a garden. Later, I learned it was because of symmetry! Our brains like patterns, and symmetry is a type of pattern. Symmetrical patterns, like flower petals, are calming.

Some bilateral symmetry is only symmetrical in one direction. You can cut this orchid symmetrically in half from top to bottom but not side to side.

What if flowers had pollen on one side of the plant and petals on the other? Where would the poor bee land?

Bees look for symmetrical flowers to find food. When they land in the middle of a flower for a snack, orange stuff called pollen sticks to their legs. As they flit from flower to flower, they spread the pollen and help more plants grow.

Ah, the garden in the winter!
It's the perfect place to
spot snowflakes.

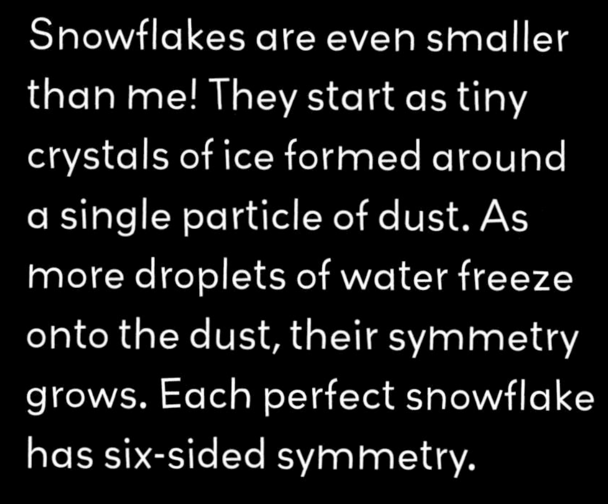

Snowflakes are even smaller than me! They start as tiny crystals of ice formed around a single particle of dust. As more droplets of water freeze onto the dust, their symmetry grows. Each perfect snowflake has six-sided symmetry.

Pretty **cool**, huh?

How many different six-sided symmetrical snowflakes can you draw? And can you say "six-sided symmetrical snowflakes" six times fast?

Time for your help! Let's make our own model gardens.

Gather a bunch of small items from around your house or outside, like coins, berries, stones, or leaves. Use your objects to create symmetrical flowers like the ones I made.

Plant your own imaginary garden!

For the next stage of the Grand Symmetry Tour,
we're going beneath the surface of . . .
. . . the ocean!

Find as many items with asymmetry, bilateral symmetry, and radial symmetry as you can!

HINT: The flounder only has two eyes.

Answer: Asymmetry: hermit crab, flounder, seaweed, purple shellfish.
Bilateral symmetry: starfish, ray, sea urchins, shell, clownfish, coral.
Rotational symmetry: starfish, sea urchins.

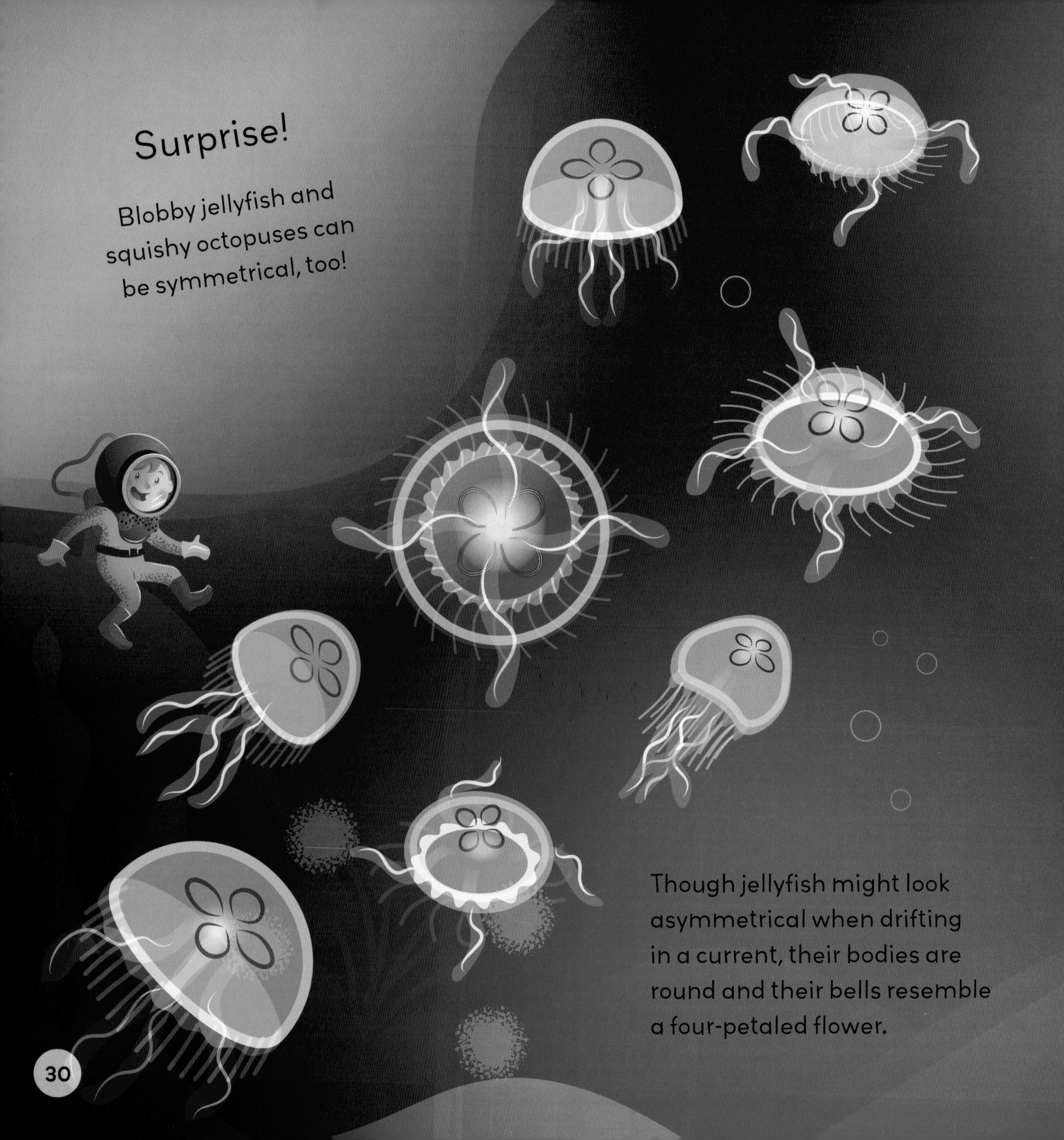

Surprise!

Blobby jellyfish and squishy octopuses can be symmetrical, too!

Though jellyfish might look asymmetrical when drifting in a current, their bodies are round and their bells resemble a four-petaled flower.

Like jellyfish, octopus often don't look symmetrical. They use their eight arms for everything: to taste, to swim, and to grab objects.

But each arm grows from a center point, and each sucker lining the arms has rotational symmetry, too. Just don't get too close when you look!

What if jellyfish and octopus bodies never changed shape? If they were totally stiff, straight, and symmetrical, could they swim?

Look out for sharks!

Oops, I meant look *at* sharks! Fish, like these speedy hammerhead sharks, swim by flexing their powerful tails back and forth. This moves them forward. Symmetrical fins on either side help them steer.

If fish fins weren't symmetrical, fish would have a hard time swimming a straight line—or in any other direction.

Symmetry helps land animals, too! What if a tiger had short legs on one side and long legs on the other?

Join in the sea creature dance party to help get us to our final stop.

Try to move across the room like . . .
An octopus, crawling across the ocean floor
A dolphin, diving in the waves
A jellyfish, bobbing through the ocean
A shark, chasing its dinner
A crab, skittering by a coral reef

Our next and final stop of the Grand Symmetry Tour is . . .

. . . the city!

Whenever I visit a city, I'm amazed at how many useful, symmetrical human-made objects there are.

Find as many human-made items with asymmetry, bilateral symmetry, and rotational symmetry as you can!

HINT: Does a car have a steering wheel on both sides?

Answer: Asymmetry: car, brown building.
Bilateral: tram, cones, Ferris wheel, tower, windows.
Rotational: Ferris wheel, manhole cover.

Wheels were invented over 5,000 years ago—and they're still *wheely* cool!

Wheels have rotational symmetry, and they're great at rotating! Wheels take us from place to place, deliver our food to the grocery store, and help build things quickly.

Wheels can be big as a house or even smaller than me. Different objects need different types of wheels.

What would happen if you mixed and matched wheels on a car?

If I had a human driver's license, instead of this ladybug one, I would love to drive on a **highway cloverleaf**. They happen where two big highways meet.

The cloverleaf lets cars switch between two big highways without stopping at a traffic light or stop sign.

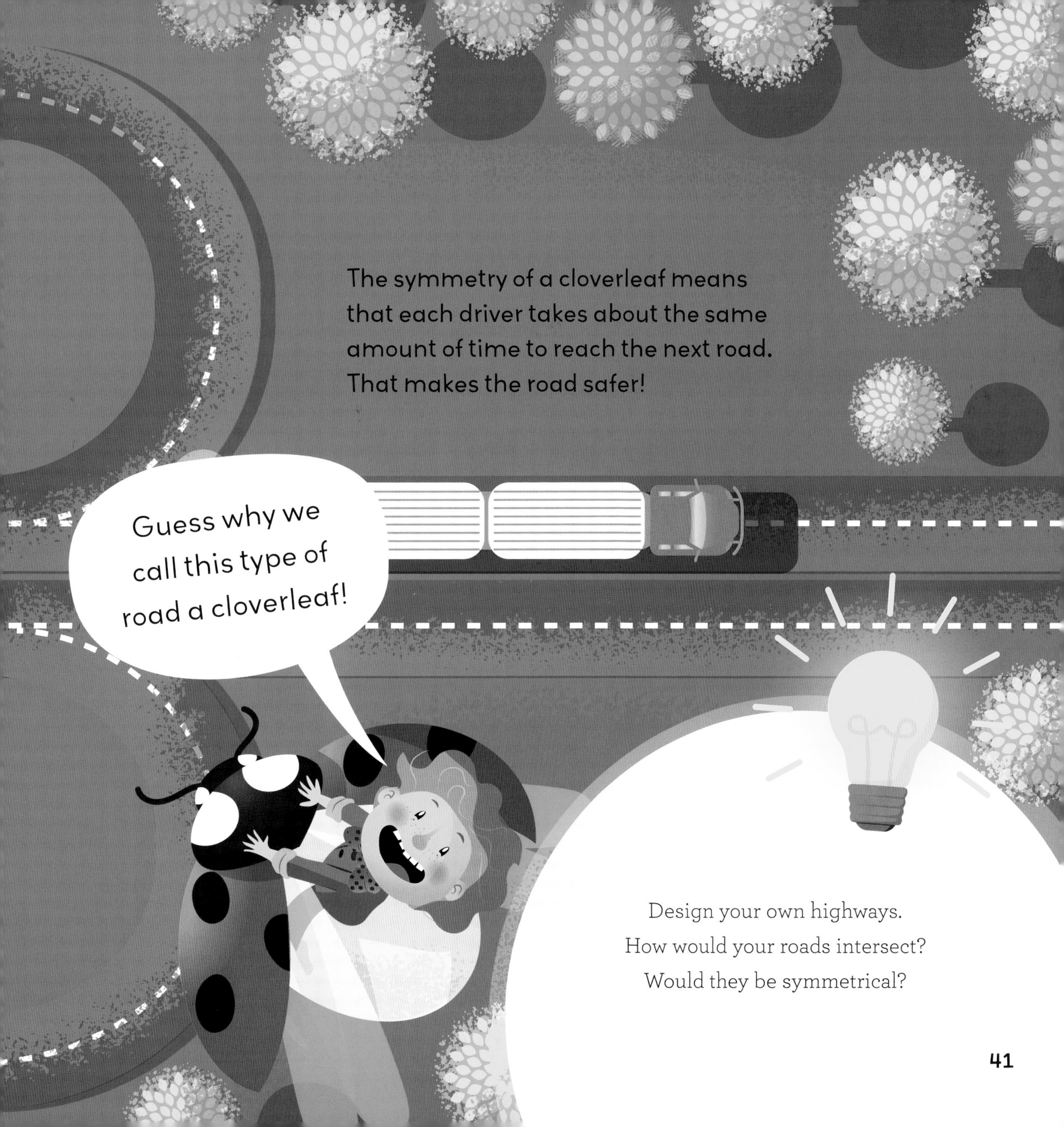

The symmetry of a cloverleaf means that each driver takes about the same amount of time to reach the next road. That makes the road safer!

Design your own highways.
How would your roads intersect?
Would they be symmetrical?

To get back to the station, we have one last challenge! Roll with me to the wheel shop.

Can you find the five sets of matching wheels on this page?

HINT: There are two sets of two wheels and three sets of four wheels.

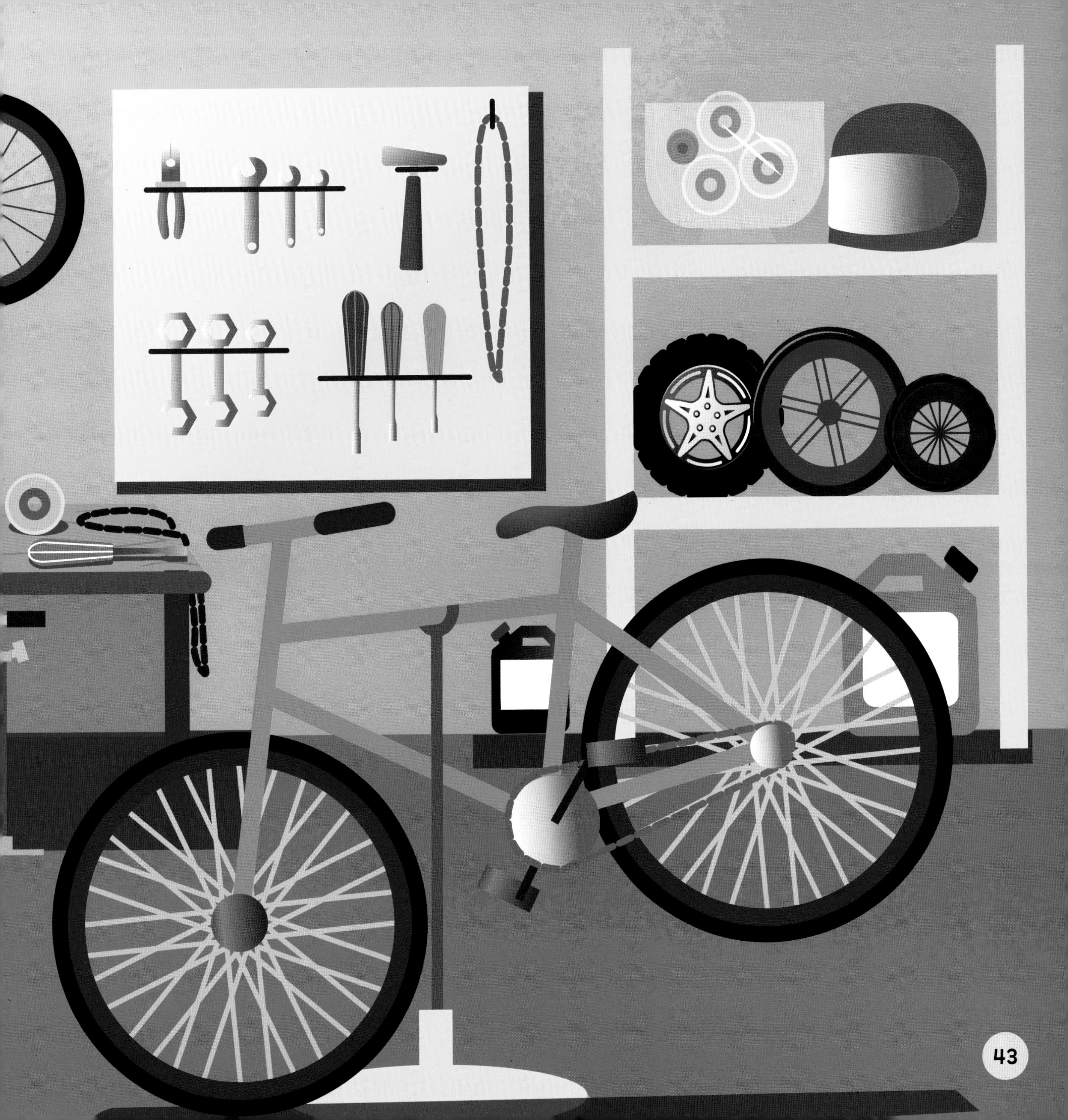

Congratulations, fellow adventurer!

We've completed the Grand Symmetry Tour!
Using your creativity and smarts,
you explored your own home,
the natural world, and even the big city.
What's *your* favorite symmetry?

Symmetrically yours,
OTTO